03/17

Classical physics

By the end of the nineteenth century many physicists believed there really wasn't much more to learn about the workings of nature and the properties of matter and radiation.

Two centuries earlier, Isaac Newton (1642–1727) had outlined, in powerful mathematical detail, what became known as *classical* or *Newtonian mechanics*. Every school-child learns about his Laws of Motion and Universal Law of Gravitation, which are used to calculate the forces that act on all objects we see around us and to explain the way those objects move – everything from falling apples to the trajectories of the Apollo moon rockets. Newton's body of work, together with the observations and careful experiments of others before him, such as Galileo (1564–1642), had set the scene for the giants of nineteenth-century physics, such as Michael Faraday (1791–1867) and James Clerk Maxwell (1831–1879), to complete the picture of *classical physics*.

This entire body of work, which took three centuries to develop, still provides us today with a description of the universe made up of solid objects obeying Newtonian mechanics, along with Maxwell's description of electromagnetic waves, fields and radiation that beautifully unified electricity, magnetism and light.

Physics, so it seemed at the time, was complete. It was expected that everything in the universe could be describable precisely, and that it was now simply a matter of dotting the 'i's and crossing the 't's. One physicist even remarked, in 1894: 'It seems probable that most of the grand underlying principles have been firmly established.'

The dawn of modern physics

In the final decade of the nineteenth century a number of mysteries and unresolved puzzles in physics were beginning to suggest that science was in crisis.

For example there was no proof as yet that matter was ultimately composed of atoms – fundamental building blocks that could not themselves be subdivided. Many physicists and chemists had begun using the idea of the existence of atoms (a rudimentary 'atomic theory') as a working assumption while in fact having no idea about their nature or properties.

There were also puzzling phenomena that could not be explained, such as the photoelectric effect, whereby light shone on metal electrodes could create electricity, as well as black-body radiation (the heat and light given off by certain non-reflecting objects) and the distinctive pattern of lines in the spectrum of light given off by each chemical element.

Adding to the excitement were three discoveries made in quick succession: first came mysterious X-rays (by the German physicist Wilhelm Röntgen in 1895), followed by the equally mysterious phenomenon of radioactivity (by Frenchman Henri Becquerel in 1896 and for which he won a Nobel Prize, with Marie and Pierre Curie), and, finally, the electron, the very first elementary particle, credited to the English physicist J. J. Thomson in 1897.

Röntgen's X-rays allowed us to see through solid matter, as though by magic, while Marie Curie earned her place in history through her work on radioactivity, becoming the first woman to receive a Nobel Prize.

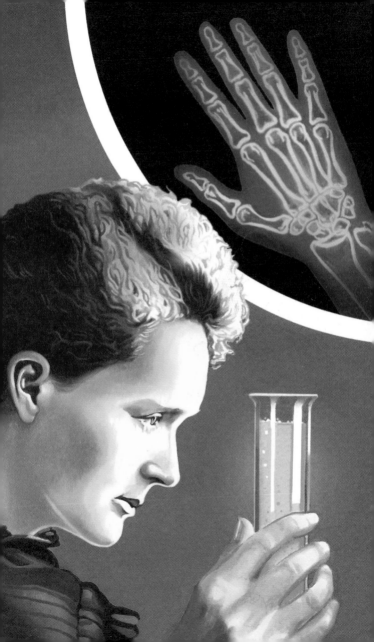

The birth of quantum theory

In 1900 the German physicist Max Planck explained the way hot bodies emit electromagnetic radiation (as heat and light) at different wavelengths. He proposed that the energy of this radiation was proportional to its frequency: the higher the frequency (or the shorter the wavelength), the higher its energy will be, and the ratio of the two quantities (energy divided by frequency) was therefore a constant number, which still bears his name: *Planck's constant*.

The idea led Planck to the conclusion that this radiation could not have just any energy – since only discrete energies were allowed (corresponding to the frequency), the radiation had to be 'lumpy', like water dripping from a tap rather than flowing continuously. It was a revolutionary proposal and completely at odds with the then prevailing ideas.

Planck's constant is a very tiny number and tells us that there is a limit to how small a lump of heat radiation can be – an indivisible piece of energy, or a 'quantum' of radiation.

This was the first indication that different rules applied at the tiniest scales. Planck was a somewhat reluctant revolutionary and was never really happy with his new theory, even though it beautifully explained the way bodies radiated heat, when all previous attempts had failed. Others would, however, take his idea and run with it.

The word *quantum* comes from the Latin *quantus*, meaning 'how great', and came into general use in physics in the first few years of the twentieth century to denote the *smallest indivisible piece*.

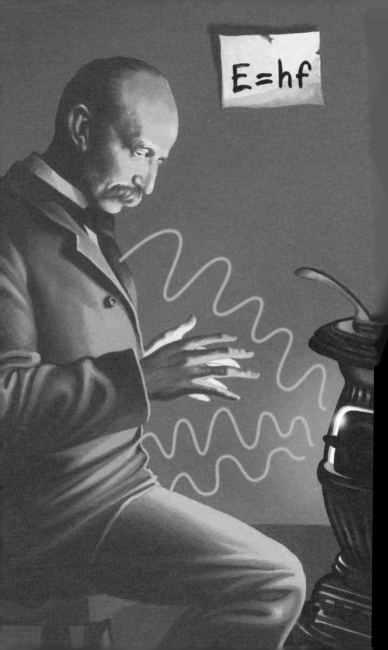

Particles of light

Planck's formula was, by and large, ignored for five years until Albert Einstein used it to explain the photoelectric effect – work for which he would win his Nobel Prize, rather than for his far more famous theory of relativity.

In the photoelectric effect, light shone on an electrically charged metal plate can knock free electrons from its surface. You might think that the energy of the released electrons would depend on the light's brightness, or intensity. Instead, their energy was found to depend on its frequency. This is an unexpected result if light is a wave, because increasing a wave's intensity (its height) *should* increase its energy. Think of water waves crashing against the seashore; they have more energy the higher they are, not the faster they break against the shore. In the photoelectric effect, high-intensity light did not result in electrons being knocked free with more energy, just *more* electrons being knocked free!

Einstein successfully explained what was happening by proposing that all electromagnetic radiation (from high-energy gamma rays and X-rays, to visible light to radio waves) is ultimately made up of tiny lumps of energy: particles we now call photons. Each electron is thus knocked out when it is hit by a single photon, the energy of which depends on its frequency. The reason we do not usually see this particle-like nature of light is due to the large number of photons involved, just as we cannot usually see individual pixels on a photograph.

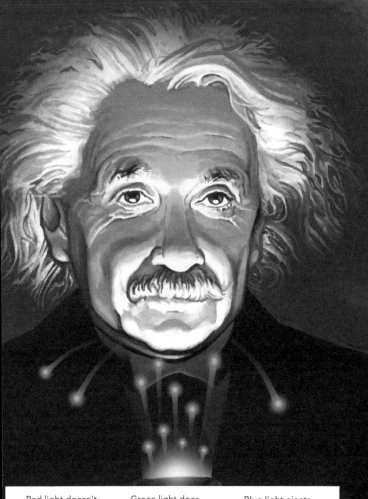

Red light doesn't eject electrons, however bright it is.

Green light does knock electrons free, however dim it is.

Blue light ejects electrons with more energy than green light, however dim it is.

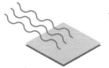

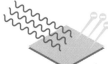

Looking inside atoms

In 1909, in Manchester, Ernest Rutherford conducted a famous experiment with Ernest Marsden and Hans Geiger into the structure of atoms.

They discovered that when a beam of alpha particles, tiny pieces of atomic nuclei given off as alpha radioactivity, was fired at a thin gold foil, the majority passed through easily, suggesting that the atoms were mostly empty space. However, a few rebounded back. Rutherford exclaimed: 'It was as though you had fired a fifteen-inch shell at a piece of tissue paper and it had bounced back and hit you.'

The only explanation was that most of an atom's mass and electric charge must be concentrated in a tiny volume in its centre, the 'atomic nucleus', which is 100,000 times smaller than the atom itself. That is, if an atom were a football stadium, its nucleus would be a garden pea on the centre spot.

But there was a problem: if all the positive charge was in this tiny nucleus while the negatively charged electrons orbited around it like planets around the Sun, why did these electrons, which are attracted to the positive charge, simply not fall into the nucleus and neutralize it? Unlike the Earth in its stable orbit around the Sun, electrically charged particles emit radiation and lose energy when forced to move in a circle, and as they do so they should, if they obeyed Newtonian mechanics, quickly spiral in towards the nucleus. But they don't – atoms are stable, or we wouldn't be here.

Rutherford observed a fluorescent screen through a microscope and saw pinpricks of light as alpha particles were deflected by a gold film.

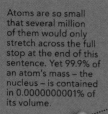

Atoms are so small that several million of them would only stretch across the full stop at the end of this sentence. Yet 99.9% of an atom's mass – the nucleus – is contained in 0.0000000001% of its volume.

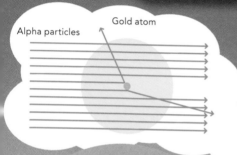

Alpha particles

Gold atom

Prevailing theory at the time predicted that the weak electrical fields in an atom would barely affect the alpha particles. Rutherford realized the scattering meant there was a tiny electrically charged centre to the atom.

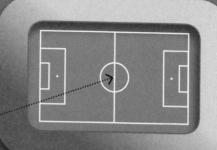

Quantizing atoms

Rutherford's problem with his solar-system model of the atom would be tackled by a Danish physicist who is ranked alongside Einstein as one of the greatest thinkers of the twentieth century. His name was Niels Bohr and he is regarded as the father of quantum mechanics.

Bohr postulated that atomic electrons lose and gain energy only according to certain 'quantum' rules. Each electron orbit is associated with a specific energy, so electrons are not free to follow just any orbit. Instead, they remain in fixed orbits, like a train on a circular track. It would later be shown that each of these orbits can accommodate only a precise number of electrons. Therefore, an electron can drop to a lower orbit only if there is space for it and only if it loses a quantum of electromagnetic energy – a photon – that equals the difference in energy between the two orbits. Likewise, it can jump to an outer orbit only if it gains a photon with the precise energy needed for the jump. When it comes to these quantum jumps, there is never any change given, only the exact energy can be exchanged.

We will soon discover, however, that the Rutherford–Bohr model of the atom is not the final word. Bohr's work on quantized electron orbits brought to an end the first phase of the quantum revolution, which is today referred to as the 'old quantum theory'.

The atomic model with electron concentric orbits (shells) that is still taught at school. The number of electrons in each is fixed. Shown are the atoms of four different elements: the lightest, hydrogen, and three of the noble gases.

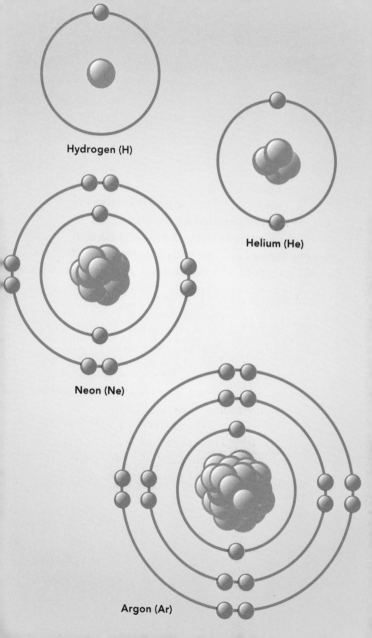

Hydrogen (H)

Helium (He)

Neon (Ne)

Argon (Ar)

Bohr creates the Copenhagen empire

In 1916 Niels Bohr returned to Denmark from Manchester, where he had helped Rutherford explain the stability of atoms. But apart from these fixed, or quantized, orbits that did not fit into the Newtonian picture, it was still assumed that Newtonian mechanics could describe the rest of the microscopic world. Electrons were thought of as tiny spheres moving along well-defined paths.

Bohr founded a new Institute for Theoretical Physics in Copenhagen in 1921 and set about gathering around him some of the greatest young geniuses in Europe, most notably Werner Heisenberg and Wolfgang Pauli. These physicists would turn the world of science upside down.

The mathematical theory of quantum mechanics – to distinguish it from classical mechanics – went far beyond simple quantization of electron orbits. It became clear that atoms were not like miniature solar systems at all and that electron orbits had to be replaced by the idea of fuzzy clouds of 'electroness'.

The theory, completed by the mid-1920s, described a world far stranger than many could accept. The Copenhagen school of thought that Bohr and others developed to describe the quantum world has shaped many areas of science and philosophy for almost a century.

On the eightieth anniversary of Bohr's birth – 7 October 1965 – his institute was officially renamed The Niels Bohr Institute. It was originally funded partly by the charitable foundation of the Carlsberg Brewery, making it . . . probably . . . the best quantum institute in the world.

NIELS BOHR INSTITUTET
1920

Wave-particle duality

In 1924 a young French nobleman called Louis de Broglie made a daring proposal: if a light wave can also behave as a stream of particles, then can moving particles of matter also be made to spread themselves out over space like waves? He suggested that every material object could be associated with a 'matter wave' that depended on its mass; the more massive a particle, the shorter its wavelength.

Experimental confirmation of the wave-like nature of electrons came in 1927, when they were shown to give rise to interference effects, just like light waves, sound waves or water waves. Just how incredible this notion is can be underlined by the famous two-slit experiment.

Matter particles, such as electrons, are fired one at a time at a screen with two narrow slits. If they obeyed the rules of common sense, then each electron that managed to get through would have necessarily gone through one slit or the other. But instead, the pattern of fringes that builds up on the screen is what is seen when spread-out waves pass through both slits simultaneously, like a single extended sea wave hitting the length of a shore. Each electron *must* be somehow behaving as a wave as it passes through, even though it hits the back screen as a particle, since the pattern on the screen is built up from individual dots where electrons have landed. If you think this is confusing then congratulations, you see the problem: it is completely crazy.

Like the particle that can go through both slits at once, the quantum skier can go round both sides of the tree. It looks impossible, but down in the quantum world, this is the equivalent of what we see happening.

Two slits

Single particle

Fluorescent detector screen showing the accumulated marks of many particles.

The pattern created suggests each particle is acting like a wave passing through both slits at once. The waves emerging from each slit overlap and interfere with each other. A particle is more likely to arrive where two wave peaks coincide on the screen – and will never appear where the waves cancel each other out.

Schrödinger and his wave equation

While Bohr and Heisenberg were developing their mathematical picture of atoms, an Austrian physicist named Erwin Schrödinger was presenting a different approach, suggesting that the entire subatomic world is made up of waves. His theory became known as wave mechanics and the famous equation that bears his name describes how such 'quantum' waves change over time.

Every student of physics and chemistry will learn about Schrödinger's equation. It would take too long to try to explain it here, so all you need to know is that it provides us with a mathematical quantity called the *wave function*. This is essentially a set of numbers that tells us everything we could possibly know about whatever it is the equation is describing, whether this is the state of a single particle, such as an electron, or an entire system of interacting particles.

Nowadays we regard Schrödinger's equation as the quantum world's equivalent of the classical equations of motion, which involve distance, speed and acceleration, but with a crucial difference. If we know where something, say a falling apple, is, and how fast it's moving at any moment in time, then we can calculate precisely when it will hit Newton on the head. But if we know the state and whereabouts of an electron now and we track how its wave function changes over time, we can only ever compute the *probability* of the electron being somewhere else in the future. We will not know for sure – we must give up on certainty.

Electron orbits are really more like clouds of probability depicting where the electron is 'most likely to be'. This is far from the schoolbook picture of a miniature solar system with electrons in circular orbits around the nucleus.

Heisenberg and the uncertainty principle

Werner Heisenberg's contribution to quantum mechanics was profound. In 1925 he formulated a new and strange kind of mathematics to describe atoms. In contrast with Schrödinger's *wave mechanics*, his far more abstract approach was called *matrix mechanics* and suggested that the spread-out nature of quantum particles wasn't a real wave at all, but an abstract mathematical entity that becomes physically tangible and solid only when we observe it.

Schrödinger, on the other hand, preferred to think of the quantum world as physically real, even if his wave function could only give us probabilities: that an electron was 'really' spread out, even when we were not looking. Heisenberg hated this, arguing that we must give up on ever being able to picture atoms accurately.

Today we have learnt to live with these two complementary ways of viewing the quantum world: Heisenberg's abstract mathematical way and Schrödinger's wavy way. Other quantum pioneers went on to show that these two seemingly incompatible approaches were, in fact, equivalent.

In 1927 Heisenberg proposed his famous uncertainty principle, which states that we can know either the location of, say, an electron, or its speed, but not both at once. This is not a result of the experimenter giving the electron an unavoidable kick through the very act of measuring its position. Rather it gives us a limit on what we can predict about the quantum world.

Explaining chemistry

In the late 1860s Dmitri Mendeleev came up with the periodic table, in which elements of similar physical and chemical properties were grouped into families. But it was only in 1925 that the Austrian prodigy Wolfgang Pauli discovered that these properties were dependent on how an element's electrons occupied its atoms' quantum orbits, or shells. Each electron has a set of numbers that define its 'quantum state'. He pointed out that no two electrons in the same atom can be in the same quantum state.

This explains why electrons don't all drop down to the lowest-energy orbit. Instead, they fill successive 'shells', the number in each governed by those quantum rules first laid out by Bohr. Once a shell is full, further electrons must occupy the next one out. The electrons in the outermost shell then govern how atoms bond together to make the huge variety of possible chemical compounds as well as explaining their physical properties, such as how well they conduct heat or electricity.

Broadly speaking, particles of matter like electrons, along with the constituents of the nucleus, protons and neutrons, which obey Pauli's rule, known as the Exclusion Principle, are called fermions. Conversely, particles of pure energy, such as photons, which do not obey this principle are called bosons. The difference, Pauli explained, is due to the way they 'spin' – not in a logical way like a basketball, but rather in a very strange quantum way that, for example, even allows for electrons to spin in both directions at once!

How the Sun shines

An intriguing yet important quantum concept is called tunnelling. It is a process that explains some of the most fundamental mechanisms in the universe.

This is when a quantum particle, such as an electron, proton or alpha particle, can hop from one side of an energy barrier to the other in a way that should be impossible. Consider trying to roll a ball up over a hill. You would need to give it enough energy to get it to the top. But in the quantum world, there would be a small probability that the ball could disappear from one side of the hill and reappear instantaneously on the other.

To visualize this, we must think of the electron not as a tiny localized particle, but as a fuzzy cloud that can seep through the barrier such that, at any given time, there is a non-zero probability that it will 'materialize' on the other side.

Tunnelling plays a vital role in sustaining our Sun, hence all life on Earth. The process that allows the Sun to shine is called thermonuclear fusion, whereby two protons (nuclei of hydrogen) fuse together to make helium and release large amounts of energy in the process. You might expect that the positive charge of the protons means they repel each other and cannot stick, just like two north poles of magnets. But thanks to their wave nature and quantum tunnelling, they can sometimes leak through their repulsive force field and get close enough together to fuse.

Our Sun is only hot enough to fuse hydrogen to form helium and a few other light elements. More massive stars can make heavier elements like lead and gold when they explode in a supernova.

Like a ball tunnelling through a hill instantly, instead of being pushed up and over it, quantum particles can sometimes appear on the other side of an energy barrier – as if by magic!

Dirac and antimatter

In a recent poll, the Englishman Paul Dirac was voted the fifth greatest physicist of all time, after Newton, Einstein, Maxwell and Galileo, which gives you some idea of the high regard in which this unsung hero is held in science.

Dirac was a shy man who was often more concerned with the elegance of his mathematics than the results of experiments. He once said:

I think it's a peculiarity of myself that I like to play about with equations, just looking for beautiful mathematical relations which maybe don't have any physical meaning at all. Sometimes they do.

He showed that the two different ways of describing the quantum world, due to Heisenberg and Schrödinger, were equivalent, and went on to modify quantum mechanics to take into account particles moving at close to the speed of light. He had to invent a new equation that now bears his name.

The Dirac equation famously predicted the existence of antimatter. Today, we know that for every matter particle there can exist its antimatter counterpart, but if a particle and its antiparticle come into contact they will annihilate in a burst of pure energy. However, the notion of an antimatter bomb is still very much science fiction, so don't worry. The process of matter-antimatter annihilation can also happen in reverse, whereby a quantum of energy, such as a photon of light, can be converted into a pair of particles, such as an electron and its antimatter partner, a positron.

Proof of antimatter: the spiral lines show tracks of an electron/positron pair being created and bent in opposite directions by a magnetic field because they have opposite electric charges.

Clash of the titans

The nature of reality described by the new quantum mechanics espoused by Niels Bohr and the Copenhagen school of thought was so strange and counterintuitive that many physicists, including Einstein himself, were unhappy with it. Things came to a head at the now famous Solvay Conference in Brussels in 1927.

For several days over the course of the week there, Einstein would present Bohr with arguments in the form of thought experiments in which he claimed to show that quantum mechanics was not the whole story and that the only way to avoid all the weirdness was to insist that it could not be complete in its current form. Each day, Bohr would go away and mull over the problem, only to return the following morning and demolish Einstein's argument. Eventually, Bohr even used Einstein's greatest contribution to science, his general theory of relativity, against him by showing that it was consistent with the predictions of Heisenberg's uncertainty principle.

History books tend to record that the Solvay Conference essentially marked the completion of the mathematical foundations of quantum mechanics. They also claim that Einstein's insistence that 'God does not play dice', referring to the unpredictability and fuzziness of the quantum world, was shown to be a forlorn hope – that we must accept the strangeness of reality at the tiniest scales.

Nevertheless, the arguments over the meaning of it all still rage on to this day.

Cats in boxes

In 1935, Erwin Schrödinger proposed one of the most famous thought experiments in science to highlight its weirdness.

He asked what would happen if a cat were shut in a box with a radioactive substance and a container of poison that would be released when the radioactive material emitted a particle.

Since quantum mechanics tells us we cannot know the moment of decay of a radioactive atom, when the box is closed we can only ascribe probabilities to the process, and thus to whether the cat is alive. Thus, we must describe the particle as having been both emitted and not emitted at the same time. We say it exists in a quantum 'superposition'.

Since the state of the cat now rests on a quantum event, it too must be both dead and alive simultaneously. We only force it to 'choose' when we open the box to look.

Is this really any different from the situation of not knowing what birthday presents you have until you unwrap them? Is it indeed just a metaphysical debate?

One sensible way of resolving the issue is to assume that such quantum superpositions exist only at the atomic scale and 'leak away' in a process called 'decoherence' that occurs when a tiny isolated quantum system is forced into contact with its surroundings. So the superposition no longer survives once we get to large objects like cats, made up of trillions of atoms, which can never be in two states at once.

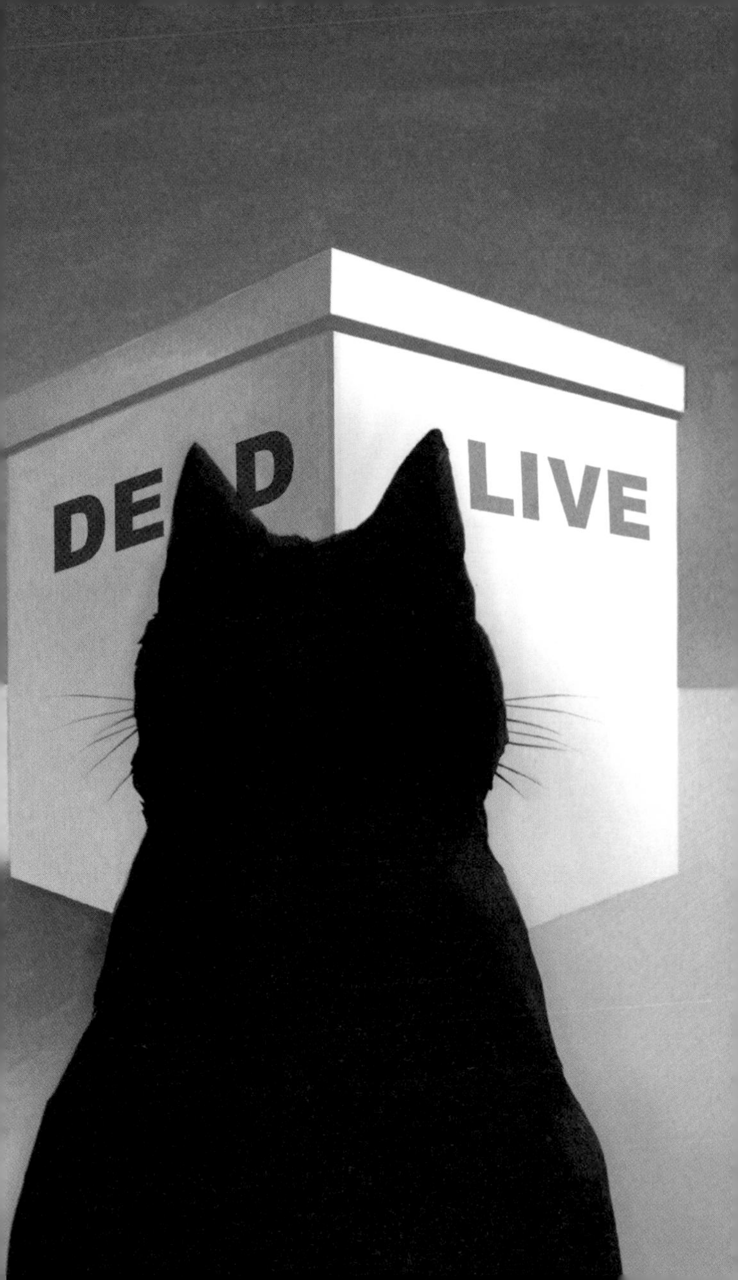
LIVE

Digging deeper

By the mid-1930s just a handful of elementary particles were known to exist. They included the proton and neutron that make up the atomic nucleus. But powerful accelerators (the most well-known example of which today is the Large Hadron Collider at CERN, near Geneva), were soon smashing these particles together at ever higher energies, creating new ones in the process.

Soon, so many new particles were discovered that a new classification scheme was required. To restore order, Murray Gell-Mann and George Zweig proposed that protons and neutrons were, in fact, not elementary at all, but were composed of tinier constituents called 'quarks'. Their hypothesis was confirmed in a series of experiments between 1967 and 1973 at the Stanford Linear Accelerator in California.

Today, we know there to be six 'flavours' of quark: 'up', 'down', 'strange', 'charm', 'top' and 'bottom'. Protons and neutrons are made of just up and down quarks. Along with quarks, another group of particles, called leptons, includes the electron and its two heavier relatives, the muon and the tau, and three types of neutrino.

The Standard Model of particle physics is like a periodic table for elementary particles. It also includes the force-carrying particles: the photon, gluon and Higgs, called bosons.

The Large Hadron Collider in CERN is hunting for an entirely new family of particles called supersymmetric particles that may or may not exist, but which would help solve a number of mysteries in physics.

The Standard Model explains much about the building blocks of the universe, but not everything; it cannot explain gravity, for instance.

There is clearly more to be dug up.

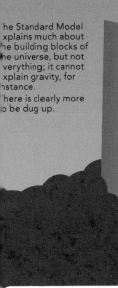

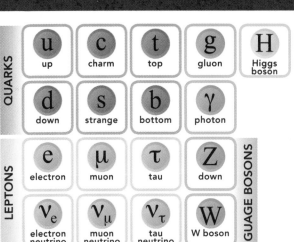

Spooky action

Possibly the strangest feature of the quantum world is the notion of entanglement – so strange that even Einstein refused to believe it, calling it 'spooky action'. This is the process whereby two separated particles remain 'connected' such that anything happening to one of them will *instantaneously* affect its partner. This is referred to as a non-local connection, which is not possible in Newtonian mechanics involving everyday objects because communication faster than the speed of light is forbidden.

But non-locality and entanglement are commonplace in the quantum world. Mathematically, this is just the extension of the idea that particles can sometimes behave like spread-out waves. If two particles come into close contact with each other they can become correlated and behave as a single entity, even if they are then moved to opposite sides of the universe. Even more incredibly, if one of them is put into a quantum superposition of being in two states at once, then the second particle will also be forced into a superposition. So, by measuring one we destroy the superposition of its remote partner, instantaneously and regardless of the distance between them.

A word of warning, though: please do not think you can appeal to quantum entanglement to explain non-scientific ideas such as telepathy. Like other quantum phenomena, it is constrained to the subatomic realm. We will see later, however, that entanglement can be used in some ingenious applications.

Quantum field theory

In the late 1940s three physicists, including the great Richard Feynman, came up with a powerful theory called *quantum electrodynamics*, or QED for short. It was a generalization of quantum mechanics and provided a new way of describing the way matter interacts with light – in fact, how all matter is held together through the electromagnetic force.

QED is an example of what is called a quantum field theory, an idea that goes back to the late 1920s with the work of Paul Dirac, who wrote a pioneering paper that combined quantum mechanics with Maxwell's theory of electromagnetism.

However, throughout the 1930s and 40s, quantum field theory was plagued by a troublesome mathematical problem: calculations always led to infinite answers when they should have been finite. The basic premise is that empty space is never truly empty, but is, in fact, a froth of particles and anti-particles continuously appearing and disappearing. However, this means that even something as straightforward as the electrical force between two electrons has to be written as an infinite series of ever-more complex processes taking place in the space between them.

The issue of the infinities was solved by QED, which is today regarded as the most accurate theory in all of physics. Feynman's approach is particularly appealing, and typical of his skills as one of the greatest scientific communicators of all time, because it makes use of powerful pictorial representations called Feynman diagrams.

But it works! Applying quantum mechanics

Quantum mechanics lies at the heart of so much of modern physics and chemistry, and plays a central role in our everyday lives, often in ways you might not realize. The rules of quantum mechanics explain how electrons arrange themselves in atoms and how atoms fit together to make molecules, and hence govern the nature of all the materials we see around us. For example, without quantum tunnelling we would not have understood how electricity is conducted in semiconductors, so we would not have created the silicon chip, and so would not have the computer or the internet. And the way electrons spit out photons of light led to the invention of the laser, which is used in all sorts of medical and industrial applications, not to mention leisure and entertainment – your DVD player reads the data using a laser.

We have quantum mechanics to thank not only for the technology behind the ubiquitous microchip, but also for another device found in electronic circuits that exploits quantum mechanics more directly: the tunnelling diode, which is used as a very fast switch in microprocessors. Quantum tunnelling also gave us nuclear power, electron microscopes and MRI scanners, which make use of quantum spin. Even the smoke detectors in your home rely on quantum tunnelling of subatomic particles.

And just consider that smartphone of yours that you've been using to google something on those rare occasions when I've lost you. It is packed with components that only do what they do because of our understanding of quantum mechanics.

Quantum 2.0

Devices that fall into the generic term of 'quantum technologies' are considered part of the second quantum revolution (or Quantum 2.0) – to distinguish them from the first quantum revolution of lasers, microchips and MRI scanners. Quantum technologies encompass a class of devices that can manipulate exotic states of matter through quantum superposition, tunnelling or entanglement.

Advances in areas such as quantum information theory, quantum electronics, quantum optics and nanotechnology are helping to develop such devices as highly accurate sensors, atomic clocks, quantum processors and secure communication instruments that use quantum cryptography. It's all very exciting.

Quantum cryptography, for example, relies on something called quantum key distribution to guarantee secure communication, because the key to encrypt and decrypt messages and information relies on pairs of quantum-entangled particles. To obtain the key, any eavesdropper or hacker would have to do the equivalent of intercepting and measuring one of the entangled particles, but in doing so would inevitably destroy their delicate quantum state, thus raising the alarm.

For now, we can make do with 'non-quantum' public key cryptosystems. The public key encryption, which is still virtually impossible to crack, means we can safely submit our credit card details online. However, if and when we invent quantum computers then public key cryptosystems would be compromised and we would then need to move to quantum encryption.

How to build a quantum computer

Quantum computers make direct use of quantum mechanics to perform operations on data, and differ from the 'classical' binary computers that we use today.

A quantum computer is based on the idea of the 'quantum bit' or qubit. In a classical computer, the basic component is the bit, which is either on or off (0 or 1). However, a qubit can exist in both states at once: a quantum superposition of 0 and 1 at the same time. By entangling many qubits together, we can achieve the ultimate in parallel processing, carrying out all computations simultaneously. Quantum computers would be able to perform certain tasks many times more quickly than even the most powerful classical computer. They are therefore expected to have many important applications.

There are currently several approaches to building a practical quantum computer. All rely on the idea of manipulating entangled superpositions of atoms, but all ultimately suffer with the same problem: how to prevent these delicate superpositions from leaking away and decohering, just like Schrödinger's cat, before the quantum task is complete.

Significant advances have been made recently, but it is debatable whether anyone has built a true quantum computer yet. However, in 2013 a consortium including NASA and Google began looking into how quantum computers might be applied in areas like artificial intelligence. It is anticipated that many applications such as this, which sound like science fiction today, will soon completely transform our world.

No one knows what a
quantum computer will
look like or how it will operate.
But we're working on it . . .

Quantum biology – a new science

While physicists and chemists have had almost a century to come to terms with the strange behaviour of the microscopic world of atoms, molecules and their constituents, biology has, by and large, not had to worry about quantum mechanics . . . until recently, that is.

The past decade has seen the emergence of the new field of quantum biology. Careful experiments are revealing that many of the weirder features of the quantum world, like tunnelling, superposition and entanglement, seem to be playing a crucial role inside living cells.

For example the way sunlight is transported through plant and bacteria cells in photosynthesis seems to rely on quantum superposition, in which the energy follows all routes at once; and enzymes, those complex molecules that help life do what it does, are getting a helping hand from quantum tunnelling. It may even be that our sense of smell, the way some animals navigate during migration or how certain mutations in DNA take place all require quantum explanations if they are to be understood.

Quantum biology is still in its infancy and many scientists remain sceptical: physicists find biology messy and complicated while many biologists find quantum mechanics too mathematical and far removed from their subject. However, while the experiments are difficult and the theoretical work complex, it's an exciting emerging field that's rich with potential.

Photons from the Sun excite electrons in the plant's cells, which then seem able to explore multiple routes simultaneously to find the most efficient pathway through the cell to where their energy can be exploited.

Photon

Reaction centre

But what does it all mean?

Niels Bohr supposedly stated, '*If you are not astonished by quantum mechanics then you haven't understood it.*' However, the mathematical part of the theory is extraordinarily powerful and accurately describes the microscopic world – it is the world it describes that is astonishing. The uniqueness of quantum mechanics as a scientific theory is that it permits the interpretation of events in many different ways.

The traditional *Copenhagen view* developed by the theory's founding fathers is not so much an interpretation as a philosophical stance. It states that we can never 'know' what is going on in the quantum world when we are not looking. All we can do is predict what we will find when we *do* look. Many physicists agree, since they don't then have to worry about what is actually going on, arguing that such issues are metaphysical and best left to philosophers.

Another is the *many worlds interpretation*, which is either the most extravagant explanation or the simplest, depending on whether you like it or not. It states that whenever we observe a particle in a superposition we do not force it to 'make up its mind', we simply observe one of the options in our universe, but our counterparts in a parallel reality see another.

There are many other interpretations, each having advantages and disadvantages. Which is the correct one, we don't yet know, since no current experiment is able to discriminate between them. We may never know the answer.

Quantum gravity – a theory of everything

So, what of the future? Physicists have long wanted to unify the four forces of nature: gravity, electromagnetism and the strong and weak nuclear forces in a single 'theory of everything'. Newton led the way by showing that the force causing the apple to fall also keeps the Earth in orbit around the Sun. Later, Maxwell showed that electric and magnetic forces are really the same thing. Today, three of the four forces have come under the umbrella of quantum mechanics. The odd one out is gravity, which is described by Einstein's general theory of relativity.

The much hyped but elusive theory of everything would bring quantum mechanics and general relativity together into a theory of *quantum gravity*. Such a theory would describe all matter and all the forces in the universe. But we are still searching for it.

Stephen Hawking took a bold step towards such a theory with a proposal requiring both quantum mechanics and general relativity. He suggested that black holes radiate quantum particles from just outside their event horizons. Such 'Hawking radiation' has yet to be observed.

Quantum mechanics can be astonishing; it can be profound and it has altered our perception of the world. We've come a very long way since those days over a century ago when quantum theory first gave the lie to the claims that science had only a few 't's to cross and 'i's to dot; but what is more exciting is that we've still got a long way to go.

Further reading

Jim Al-Khalili *Quantum: A Guide for the Perplexed* (Weidenfeld & Nicolson, 2012)

Jim Al-Khalili and Johnjoe McFadden *Life on the Edge: The Coming of Age of Quantum Biology* (Black Swan, 2015)

Jim Baggott *The Quantum Story: A History in 40 Moments* (Oxford University Press, 2011)

Jon Butterworth *Smashing Physics* (Headline, 2015)

Sean Carroll *The Particle at the End of the Universe* (Oneworld Publications, 2013)

Brian Cox and Jeff Forshaw *The Quantum Universe: Everything that Can Happen Does Happen* (Penguin, 2012)

Richard P. Feynman and A. Zee *QED: The Strange Theory of Light and Matter* (Princeton University Press, 2006)

John Gribbin *Computing with Quantum Cats: From Colossus to Qubits* (Bantam Press, 2014)

Steven Holzner *Quantum Physics for Dummies* (John Wiley & Sons, 2009)

J. P. McEvoy and Oscar Zarate *Introducing Quantum Theory: A Graphic Guide* (Icon Books, 2003)

Chad Orzel *How to Teach Quantum Physics to Your Dog* (Oneworld Publications, 2010)

Carlo Rovelli *Reality Is Not What It Seems: The Journey to Quantum Gravity* (Allen Lane, 2016)